Saskatchewan *Wild*

Saskatchewan *Wild*

A WILDLIFE PHOTOGRAPHIC JOURNEY

Robin and Arlene Karpan

PARKLAND
PUBLISHING

Saskatoon

Published in Canada in 2010 by

Parkland Publishing
501 Mount Allison Place
Saskatoon, Saskatchewan
Canada S7H 4A9

306-242-7731
info@parklandpublishing.com
www.parklandpublishing.com

Printed in Canada by Friesens.

Library and Archives Canada Cataloguing in Publication

Karpan, Robin
Saskatchewan wild : a wildlife photographic journey / Robin and Arlene Karpan.

ISBN 978-0-9809419-1-3

1. Animals--Saskatchewan--Pictorial works. I. Karpan, Arlene II. Title.

QL221.S3K37 2010 591.97124 C2010-900550-3

FIRST PAGE: *Sandhill Cranes, Luck Lake Heritage Marsh.*
PAGE 2: *Red-necked Grebe.*

Mixed Sources
Cert no. SW-COC-001271
© 1996 FSC
FSC

Acknowledgements

Many people and organizations helped us in our research for this book. We acknowledge the valuable assistance we received from Saskatchewan Provincial Parks, and from Clem Millar of Chaplin Tourism. We thank those who helped us verify the identity of plants and animals, as well as those who made many useful suggestions and provided feedback on the text and photos including Frank Roy, Chuck Deschamps of Ducks Unlimited Canada, Joe Milligan, Anna Leighton, Melody Nagel-Hisey of Cypress Hills Interprovincial Park, Lori Wilson, Phil Taylor, formerly of the Canadian Wildlife Service, among others. The interpretation of comments provided by these and other people are entirely our own, and errors that may have crept into this book are also entirely our own.

Parkland Publishing acknowledges the assistance of the Creative Industry Growth and Sustainability program, made possible through funding provided to the Saskatchewan Arts Board by the Government of Saskatchewan through the Ministry of Tourism, Parks, Culture and Sport.

Contents

Introduction

Waterfowl lift-off, Foam Lake Heritage Marsh.

We wait quietly near the shore of the shallow lake, feeling the crispness of the early October morning. A warm gold and crimson glow bathes the sky as the sun peeks over the eastern horizon. Suddenly, an explosive thrust of wings breaks the peaceful dawn as Snow Geese in the tens of thousands lift off the water in unison, almost blackening the sky, their loud "whouk, whouk" calls adding to the deafening din. The breathtaking scene ranks among nature's great spectacles, yet it is one that we easily find in Saskatchewan.

Saskatchewan abounds with wonderful opportunities to see and enjoy wildlife—some 350 different birds, an array of fascinating creatures great and small, wildflowers and plants in a kaleidoscope of colours. Much of this diversity is due to our varied habitats including boreal forest, Canadian shield, wild rivers, some 100,000 lakes, badlands, aspen parklands, Canada's largest sand dunes, and some of North America's largest expanses of native grasslands.

Our location on the Central North American Flyway gives us easy access to sites where migrating waterfowl stage in the hundreds of thousands. For nesting waterfowl, Saskatchewan has long been famous as "North America's Duck Factory", raising a quarter of the continent's ducklings. Saskatchewan is also home to half of Canada's pelicans, half of Canada's designated sites in the Western Hemisphere Shorebird Reserve Network, the world's largest inland concentration of Bald Eagles, white-tailed deer with the largest antlers in North America, some of the last remaining sage-grouse in Canada, the only black-tailed prairie dogs in Canada, and a quarter of the world's endangered Piping Plovers.

But appreciating and enjoying wildlife is much more than finding the biggest, best or rarest. Everyday encounters can be just as rewarding—hearing a meadowlark announce the arrival of spring, strolling through a meadow carpeted in wildflowers, or catching a brilliant azure flash from a flying Mountain Bluebird. It's hard to imagine that some wildlife that we almost take for granted now, such as the pelican, pronghorn, and even the ubiquitous Canada Goose, have recovered from once being on the brink. Will today's species at risk such as Greater Sage-Grouse, Piping Plover, or Burrowing Owl be as lucky?

The story of our wildlife is both good news and bad news—we are blessed with so much, but habitat necessary to sustain wildlife continues to be lost in the name of progress. Southern Saskatchewan already has some of the most altered landscapes in Canada, so it becomes critical to protect what's left of our natural areas. Northern Saskatchewan fares better, with vast areas of pristine wilderness still remaining, although "development" is encroaching on wild places at a frightening speed.

Saskatchewan Wild is a personal photographic journey, a taste of the many wildlife adventures possible in Saskatchewan. It is both a celebration of our wildlife and a reminder of what we might lose if we aren't careful. In our over-urbanized, over-civilized world, it becomes increasingly important not to lose our connection to the natural world. We hope that this book can play a role in fostering a greater appreciation of our wild creatures and wild places, and that by understanding and valuing what we have, we can better protect them.

The spiky flower-head of water smartweed, a plant found in shallow water in lakes, ponds, or creeks.

North America's Duck Factory

Saskatchewan is duck heaven—the famed "Duck Factory" where one in four of the continent's ducklings is raised. Recent counts put North America's duck numbers at around 40 million; that's a lot of quacking. For species such as Mallards, Canvasbacks and Redheads, Saskatchewan usually has more ducks than any other province or state, thanks to the proliferation of ponds, potholes and wetlands that ducks like and need.

In 2009, for example, the number of May ponds in the prairie and parkland regions of Canada was estimated at over 3.5 million, more than 60% of which were in Saskatchewan. Since duck populations fluctuate with available water, drought can cause numbers to plummet. While drought is a temporary problem, draining wetlands is disastrous for ducks and other waterfowl, resulting in permanent losses. It's been estimated that wetland loss has been at least 40% province-wide since Saskatchewan was settled, with the loss as high as 90% in some areas.

The best thing about watching ducks is that we find them practically everywhere. Hotspots abound such as Ducks Unlimited Canada projects, national and provincial parks, heritage marshes, or national wildlife areas. But in spring it's often a simple matter of going for a drive and stopping at some of the countless wetlands near the road when ducks are decked out in their finery, and may be preoccupied with colourful mating rituals.

OPPOSITE: *Female Bufflehead.* ABOVE: *Saskatchewan's waterfowl thrives in wetlands and marshes such as Kaposvar Creek near Esterhazy.*

Bulbous Beauties

The Bufflehead duck's odd name was derived from "buffalo head" because the male's bulbous head seems out of proportion to its small body. At first glance, the male is a striking black and white, but when he turns his head just right, the neck plumage shows a dazzling mix of metallic-looking purple, violet and green. Nicknamed the Spirit Duck, this graceful swimmer often glides across the water with little apparent body motion, barely causing a ripple, then disappears below the surface in the blink of an eye. The Bufflehead is North America's smallest diving duck and, according to the Canadian Wildlife Service, one of the scarcest. Fortunately, in parts of Saskatchewan's woodlands, it is fairly common.

Our farm in the aspen parkland of the Thickwood Hills is blessed with exceptional waterfowl habitat. Of all the birdlife that thrives here, it's the diminutive Bufflehead that has a special place for us. For seven consecutive years, a Bufflehead hen nested in a tree cavity just a few metres from our back door. While we can't say with certainty that it was always the same duck, her changing behavior over the years suggests that it was.

To begin with, the mother duck seemed skittish. When she sat in the tree with her head out of the nest hole, the first sign of us would send her popping back inside or flying away. In early evening, she often returned to the nest just when we were outside enjoying a barbecue. If she saw us, she would do a quick fly-by, often accompanied by the male: our cue to move. At first we would have to be almost out of sight before she would fly into the nest. Later, she was comfortable if we were only a short distance away. Later still, it didn't seem to matter. We could be sitting close by and suddenly there would be a brief swoop as this fast and graceful flyer made the most inelegant plop onto the edge of the nest then wiggled in as if she were an oversized cork being stuffed into a bottle.

Also in later years, when she rested at the entrance to the tree hole, she didn't seem at all bothered as we went about our business.

A few times we were fortunate to be around, usually the second week of June, when the ducklings hatched and popped out of the nest, bouncing on the grass like tiny fuzzy tennis balls. The mother would lead them to the nearest water, about 200 metres away. Sometimes she headed out before all the ducklings had left the nest. When this happened, we did our best to "herd" the late-comers in the right direction, staying close by across an open stretch while hungry ravens flew overhead. One little peeper was determined to go in the opposite direction, no matter what. Finally, we picked it up and carried it to the water's edge, where the mother immediately accepted it.

All good things eventually come to an end. The poor tree died, then the top two-thirds broke off in a wind storm. But the remaining three metres or so continued to be prime real estate. Besides the hole claimed by the Bufflehead, another hole was home to Tree Swallows almost every year, and a third was often used by a Hairy Woodpecker for winter shelter. We kept propping up this bird condominium as long as possible, but it eventually disintegrated. In the meantime, we put up nesting boxes closer to the water to help keep the bulbous beauties around.

OPPOSITE TOP: *Female Bufflehead in nest entrance.*
OPPOSITE BOTTOM: *Female Bufflehead and ducklings.*
RIGHT: *Male Bufflehead in breeding plumage.*

Fast Fliers, Deep Divers

One of our bigger ducks, the Canvasback has a greyish-white back, and a long, gracefully-sloped forehead and black bill, giving it a regal appearance. This fast flier can reach speeds over 100 km/hour, and dives deeply in search of food. In Saskatchewan, the Canvasback is most common in the parklands and northern parts of the grasslands where it prefers deeper marshes, larger permanent sloughs, and bays on lakes. Canvasbacks are not plentiful, recently averaging just over a half million in North America, with roughly a third nesting in Saskatchewan.

OPPOSITE: *Male Canvasback.*
ABOVE: *Female Canvasback and ducklings.*

Naughty Nesters

The Redhead is the most appropriately named duck around. The male's vivid rufous-red head provides a splash of colour in a spring marsh. In the magical light of early morning or late evening, the colour takes on an almost unreal intensity. This beautiful duck does have one less than admirable trait—it is infamous for nest parasitism. The female often lays eggs in the nest of another duck, usually another Redhead or a Canvasback. The Redhead's main breeding range encompasses the prairie marshes of Canada and the northern United States. When this region has water, Redheads thrive. Recent estimates during relatively wet years put the population at a bit over a million, up nearly two-thirds from their long-term average. Over a quarter of all Redheads nest in Saskatchewan. They prefer deeper wetlands where it is easier to dive for food.

LEFT: *Male and female Redheads.*
ABOVE: *Male Redhead.*

Waterfowl-rich potholes and small lakes in the Thickwood Hills northwest of Blaine Lake. Although these scenic aspen parkland uplands still have more natural areas than most parts of the agricultural belt, we see more and more trees being bulldozed every year.

Colourful Quackers

"If it quacks like a duck, it must be duck." That old saying seems tailor-made for the Mallard. While different ducks make different sounds, the Mallard's loud and distinct "quack, quack!" is what many people consider the quintessential duck call. The handsome male sports a golden bill on a metallic green head, a chestnut breast, and a formal-looking white collar or "necklace". The most obvious colour on the mottled female is a patch of iridescent blue on the wing, framed by white bars.

Mallards are the most abundant and widespread duck in North America. Of the eight million or so found on the continent, close to a quarter breed in Saskatchewan. Mallards are distributed throughout the province, especially around shallow wetlands. While mostly a summer resident, some Mallards find the Saskatchewan winter so amenable that they hang out here year round. Where there is open water at reservoirs, such as below Gardiner Dam, wintering Mallard numbers have been as high as 8,000.

OPPOSITE TOP AND LEFT: *Male Mallard.*
OPPOSITE BOTTOM: *Mallard duckling.*
ABOVE: *Female Mallard.*

Tiny Teals

About half the size of a Mallard, the Blue-winged Teal is an expert flier, famous for quick aerial manoeuvres, often skimming the surface of the water. It is named for the blue patches on its forewings, although these aren't always obvious. The male's most distinguishing feature in breeding plumage is his plum-tinged head with white crescents on each cheek.

These are our most well-travelled ducks, wintering as far south as Central America and northern South America, with the occasional straggler even showing up in Chile. With this twice-yearly marathon migration, Blue-wings tend to be one of the later ducks to show up in spring and among the first to leave in fall.

Favouring small sloughs and ponds throughout the parklands and grasslands, Blue-winged Teals are plentiful in Saskatchewan. Provincial estimates have been around two million some years, often accounting for a quarter or more of the North American population. But since these ducks prefer small potholes, they are sensitive to drought, their numbers fluctuating significantly with the availability of water.

Even smaller, Green-winged Teals prefer the treed wetlands of the parkland and forest, and are less common on prairie potholes. They are adept at hiding their nests, often concealing them under thick brush. The smallest of the dabbling ducks, the birds are named for their green wing bars. In breeding plumage, an even more conspicuous marking is the large green head patch on the male's reddish-brown head, and a spotted breast with a pinkish tinge. His call is more like a whistle than a quack.

OPPOSITE: *Female and male Blue-winged Teals.*
ABOVE: *Male Blue-winged Teal.*
RIGHT: *Male Green-winged Teal.*

Forever Blowing Bubbles

The male Ruddy Duck is both striking and odd-looking, with a stiff, upright spiky tail, chestnut-coloured plumage, and a thick neck that seems out of proportion to the rest of its body. But what really makes the male Ruddy stand out is his brilliant sky-blue bill.

Ruddies inhabit wetlands through much of the southern half of Saskatchewan, but they seem most at home in the small lakes and potholes of the parklands. In our neck of the woods in the Thickwood Hills, it seems that almost every body of water has its own Ruddies brightening the wetlands with flashy blue bills.

When it comes to courtship displays, the male Ruddy is more imaginative than most ducks—he uses bubbles as part of the act. He inflates an air-sac in that thick neck then pumps his bill up and down, whacking the underside of the bill against the inflated neck. This not only produces a drumming sound, but also creates an impressive array of bubbles. He ends the performance by opening his bill and making a belching sound as air is released from the air sac. What female could possibly resist?

The amorous behavior of Ruddy Ducks once had an unforeseen consequence—it got Farley Mowat fired from his first writing job. In his book *Born Naked*, Farley recounts his years growing up in Saskatoon, where he wrote a nature column for the local newspaper. Everything was going fine until he did a column on how Ruddy Ducks mate. It seems that the explicit sexual detail he included was too much for a proper newspaper of the day. Complaints were made and Farley's column was axed.

OPPOSITE AND RIGHT: *Male Ruddy Duck making bubbles in courtship display.*

ABOVE AND BOTTOM RIGHT: *Female Northern Shoveler.*
TOP RIGHT: *Male Northern Shoveler dabbling for food.*
OPPOSITE: *Male Northern Shoveler.*

"Spoonies"

The Northern Shoveler is famous for its large, broad spoon-like bill, lined with what at first glance look like teeth, but are actually comb-like projections used to strain food it picks up from dabbling in sloughs. The male's iridescent green head somewhat resembles that of the Mallard, except that the shoveler has that distinctive "Daffy Duck" bill and yellow eyes that make him appear wide awake. Like most other ducks, the female shoveler is more drably coloured, except for her bold orange bill that is much more showy than the male's black bill.

The Northern Shoveler, sometimes referred to informally as a "spoonie", is common in much of central, southern, and a few parts of northern Saskatchewan, with its prime habitat in the parklands. It favours shallow sloughs with submerged vegetation. Since shovelers are some of the later ducks to breed and nest, we often have the chance to see their courtship rituals which involve a lot of chin lifting, with mates vigorously pumping their heads up and down.

Scaup and Ring-necks

Scaup have numbered over four million in North America in recent years. Unfortunately, their numbers have also fallen significantly over time. In Saskatchewan, Lesser Scaup are much more plentiful than their slightly larger cousins, Greater Scaup, and are found in most parts of the province, especially in the parklands and boreal forest. Sometimes referred to as "little blue bills", male Lesser Scaup have pale blue bills, and heads with a subtle purple tinge that sets off yellow eyes.

The Ring-necked Duck at first glance looks similar to a scaup—dark at the front and back, and lighter in the middle. While the proper name is Ring-necked, it is referred to informally as ring-billed, which seems a more appropriate name. Although it does have a neck ring, it is not always obvious. More apparent are the two white rings on the bill of the breeding male, while the female has a single faint ring. Another distinguishing feature is its peaked or slightly pointed head. In Saskatchewan, Ring-necks tend to favour parkland and northern marshes and bogs over prairie wetlands.

OPPOSITE TOP RIGHT AND LEFT: *Lesser Scaup.*
OPPOSITE LEFT: *Female and male Ring-necked Ducks.*
RIGHT: *Male Ring-necked Duck.*

Debonair Dabbler

Considered by many to be our most beautiful duck, the resplendent male Wood Duck in breeding plumage has a green head with white stripes ending in a feathery mane, bridle-like neck stripes, and a bill with splashes of red, white and yellow. With sharp claws to perch or grab on to trees, these small ducks are cavity nesters, using holes in trees or nest boxes. Wood Ducks are not at all common in Saskatchewan, with the southeast part of the province having most of the few that nest here. One place to look for them is Regina's Wascana Lake.

Male Wood Ducks.

Pond Poachers

American Wigeons nest throughout Saskatchewan where they seek out shallow lakes and sloughs surrounded by dry meadows where they can forage for tender plant shoots. They have a rounded mottled head with dark eye patches that make it look like they got into a fight and came away with black eyes. Males have a white crown which has been likened to the bald head of a man, earning this bird the nickname "baldpate". Wigeons' strong bills enable them to pluck vegetation from the bottom of ponds. They also have a reputation as poachers, snatching food that other ducks bring to the surface. Since they can walk easier than most ducks, they spend a lot of time grazing on shore. Though American Wigeons are widespread, their numbers have fallen over time, with the biggest decrease on the Canadian prairies.

Male American Wigeon.

Gone Fishin'

Common Mergansers are the largest and most abundant mergansers in Saskatchewan. Roughly the size of Mallards, but more sleek and elongated, mergansers are specially adapted to swim under water where they catch fish and other aquatic creatures. Their slim bills have tooth-like serrations that help them hang on to wet and slippery prey.

The male has a mostly white body topped with a dark back, green head, and red bill. Unlike many ducks, the female is quite showy as well, especially her rusty-red head backed by a crest that looks like hair blowing in the breeze. The male doesn't have a crest.

Fondness for fish makes them a bird of the north, where we often see them flying straight as an arrow, low over the water. One of the best places to see mergansers is Prince Albert National Park, where they frequent the shorelines and tend to be reasonably tolerant of people.

Female Common Merganser.

Giants of the Marsh

These waterfowl giants with long graceful necks are our most common and familiar goose, found throughout the province. In southern Saskatchewan, if there's a body of water, chances are we'll find Canada Geese. They arrive and start nesting early in spring, and stay late in fall. It's not unusual to see huge flocks in November, some of these being Arctic-nesting geese that stage here on their way south. The hardiest ever hang around throughout the winter. With so many geese everywhere we look, it's hard to believe that they came close to disappearing in the 1930s due to drought and over-hunting. Today, there's seldom a problem finding Canada Geese, their most famous haunt being Regina's Wascana Park where in places you have to carefully tip-toe between goose droppings.

Canada Goose and goslings.

Grebes

Grebes seem duck-like, but look closer and they are significantly different. Grebes have thin pointed bills compared to ducks' flat bills. With relatively small wings for their size, they often have to pitter-patter across the water to muster enough momentum to fly. As a general rule, grebes are less skittish around people, making it easier for us to stop for a closer look.

One of our bigger grebes, the Red-necked Grebe is fairly common in the parklands and in many parts of the north. While it is less widespread in southern Saskatchewan, it does nest in several locales such as the Cypress Hills, Last Mountain Lake, and Regina's Wascana Lake. Named for its chestnut-red neck, the Red-neck favours larger water bodies where it can make full use of its diving abilities. It is also very vocal, with a loud boisterous call that sometimes sounds like laughing.

Eared and Horned Grebes, found mostly in the southern half of Saskatchewan, are both small birds with haunting red eyes set off by yellow patches on the sides of their heads. The Horned Grebe's tufts are thicker and more solid looking, reminiscent of horns, while the yellow patches on the Eared Grebe are more wispy. Another noticeable difference is the Horned Grebe's rufous red neck; the Eared Grebe's neck is black. Eared Grebes nest in colonies and prefer more open water, while Horned Grebes seek smaller protected wetlands.

TOP LEFT: *Eared Grebe.*
BOTTOM LEFT: *Horned Grebe turning eggs on its nest.*
OPPOSITE TOP: *Red-necked Grebe.*
OPPOSITE BOTTOM: *Horned Grebes.*

Wetland Busybodies

In many Saskatchewan wetlands where vegetation grows in shallow water, chances are that American Coots will be the most common water bird around. Sometimes called mud hens, they are seldom still, constantly fussing about their floating nests while cackling and squawking, or splashing their wings to chase away competitors. The rather drab grey and black adults have just a bit of colour on their white bills. Chicks, on the other hand, sport orange-red bills, wild downy head feathers that make every day a bad hair day, and a face that only a mother could love. When chicks are small, coots are some of the most entertaining birds in the marsh; adults dabble steadily for food to bring to the chicks, which always pester them for even more.

OPPOSITE AND ABOVE: *American Coot and chick.*

Vanishing Act

We hear these marsh "slough pumps" more than we see them. On a quiet spring evening, a slow repetitive "bloonk, bloonk" comes from deep inside the reeds, as if someone is steadily pumping an old-fashioned water pump. American Bitterns are notoriously secretive, but when we're lucky enough to spot one, we find an attractive though odd-looking bird—stocky, with legs that seem short for a large wading bird, and a long neck that won't quit. Walking extremely slowly and carefully, they always seem to be sneaking up on something. They are so well camouflaged that we likely walk right by them without ever noticing. A surprised bittern will usually stop dead in its tracks and stretch its head skyward, so its brown-streaked breast blends even more into the vegetation. We came across this long-necked beauty (right) while driving through the National Wildlife Area at Last Mountain Lake. Although it had wandered outside the cover of vegetation, it immediately struck its "you can't see me" pose and stayed absolutely still, though we were little more than a couple metres away.

ABOVE: *American Bittern feeding in a marsh.*
RIGHT: *American Bittern in its "you can't see me" pose.*

Blackbirds of the Marsh

Almost anywhere in Saskatchewan that we find a marsh, chances are that we'll find Red-winged Blackbirds. Not only are they common here, but they are also considered one of the most abundant birds in North America. These early spring arrivals waste no time staking out their territory, the males clinging to the tops of cattails, showing off their red and yellow shoulder patches, and squawking a hoarse "konk-ke-ree".

The Yellow-headed Blackbird is even more showy, the male's brilliant golden head contrasting with its black body. These blackbirds are more common in southern Saskatchewan where they prefer larger wetlands and deeper water than their Red-winged counterparts. Where the two blackbirds are found together, Yellow-heads tend to dominate the marsh, pushing the Red-wings to the edges. Yellow-headed Blackbirds have an unfortunate reputation—the worst-sounding song of practically any bird. Their bizarre raspy call has been likened to everything from a honking gargle to a rusty hinge, and even to someone being strangled. But who are we to judge? Females seem to like the songs just fine.

TOP: *Male Red-winged Blackbird.*
BOTTOM: *Male Yellow-headed Blackbird.*

Call of the Loon

The haunting call of the loon captures the spirit of the northern wilderness. There's nothing quite like camping beside a remote river or lake and hearing that plaintive cry on a quiet evening. The Common Loon has four call variations: a sweet-sounding wail, a laugh-like tremolo, a yodel that males use to defend their territories, and a short hoot that family members often use to keep in contact with one another.

It's always special to see loons at close quarters, with their elaborate checkering and spotted patterns, glossy black head, and piercing red eye. Occasionally loons will "stand" on the water with wings extended to stretch, to dry their wings after diving, or in an aggressive stance. These are powerful swimmers and divers that prey on fish, frogs or leeches. This agility and gracefulness doesn't carry over to land where loons walk awkwardly because their feet are set so far back on their bodies. To avoid walking far, they usually nest right at the water's edge to make it easier to slip into and out of the water. Consequently, shoreline "developments" are generally not good news to loons.

In Saskatchewan, loons nest throughout our fish-rich northern waterways. Some famous loon locales include Anglin Lake which has a high density of nesting birds, along with nearby lakes in Prince Albert National Park.

Common Loon "standing" on the water.

Sentries of the Shore

As the name suggests, shorebirds spend much of their time near water's edge, wading in the shallows of marshes, ponds, and lakes, or scouring beaches and mudflats for insects, invertebrates, and small crustaceans.

We find over 30 different shorebirds in Saskatchewan. Many nest here including the American Avocet, Marbled Godwit, Willet, Wilson's Phalarope, and the ubiquitous Killdeer. While those species are common, Saskatchewan is also home to the world's largest nesting population of Piping Plover, one of Canada's most endangered birds.

Equally important is Saskatchewan's critical role for migrating shorebirds such as Hudsonian Godwits, Sanderlings, dowitchers, various sandpipers, and others that stop here to feed and rest on their way to and from Arctic nesting grounds. Some of these birds travel phenomenal distances between South America and the Arctic, with only a small number of "stepping stones" along the way. To put into perspective just how important these stopover feeding sites are, the Western Hemisphere Shorebird Reserve Network has designated the most important shorebird sites throughout the Americas. Three of the six sites in Canada are in Saskatchewan: the Quill Lakes, Last Mountain Lake, and the Chaplin Lake/Old Wives Lake/Reed Lake complex, each with large expanses of shallow water and mudflats offering a shorebird buffet. Shorebirds are not restricted to these internationally famous sites. Places such as the Blaine Lakes, Manito Lake, Pelican Lake, Luck Lake, Redberry Lake, and many others often host significant numbers of nesting or migrant shorebirds.

American Avocets at sunset.

Shall We Dance

The Saskatoon Nature Society adopted the American Avocet as its emblem for good reason. With slender stilt-like legs, an elongated, slightly upturned bill, and lustrous cinnamon head and neck, avocets rank among Saskatchewan's most elegant birds. Common summer residents through much of southern Saskatchewan, avocets nest near wetlands, shallow lakes, and even temporary sloughs. They especially like saline lakes with mudflats such as Chaplin Lake or the Quill Lakes. With long thin legs tailor-made for wading, they forage by walking in the shallows while sweeping their bills through the water to find aquatic insects and crustaceans.

Mating rituals are quite a show. After walking around each other in the shallow water, the female takes the initiative by lowering her head to just above the water in a "come-hither" invitation. But the most elaborate part of the ceremony comes immediately after mating, when the couple does a circular pirouette dance with their necks stretched close together, at times crossing their bills.

Saskatchewan abounds with places to see avocets. Among our favourites is Luck Lake where we can often watch and photograph them up close while driving on the dyke road that crosses the shallow lake. They are fairly tolerant and tend to simply go about their business if we stay in our car. But get too close to a nest and it's a different story. These fearless protectors will attack predators such as hawks. People who stray too close will at the least get a loud scolding, or worse, a dive-bombing.

OPPOSITE: *American Avocets.*
TOP LEFT: *American Avocet giving a scolding.*
BOTTOM LEFT: *Well-camouflaged avocet eggs in nest.*
TOP AND BOTTOM RIGHT: *Mating dance.*

Abundant and Rare

The Killdeer is our most common resident shorebird, while the Piping Plover is our most endangered. Every region of the province is home to the Killdeer, except the extreme northeast. This noisy bird (the second part of its Latin name is *vociferus*) is named for its loud call "kill-dee, kill-dee". It is especially adept at the broken wing act. At the first sign of trouble, this well-camouflaged bird sneaks away from the nest, then at a safe distance, distracts predators by screaming and pretending to be wounded easy prey.

The Killdeer's smaller, quieter cousin, the Piping Plover, is identified by the single black ring below its neck, and its peeping or piping sound. It blends in well with the sandy and pebbly shoreline it prefers. An estimated two-thirds of Canada's and one-quarter of the world's population nests in Saskatchewan, with some of the highest numbers at Lakes Diefenbaker, Chaplin, and Big Quill. Reasons for its troubles are many including drought, habitat loss, and predation. An added complication is that one important nesting area, Lake Diefenbaker, is a "managed" reservoir often filled in late spring after eggs are laid, flooding plovers' shoreline nests.

OPPOSITE: *Killdeer.*
ABOVE: *Piping Plover at Chaplin Lake. This plover has leg bands indicating that it has been banded as part of a research project to monitor these rare birds.*
RIGHT: *Marbled Godwit.*

Marbled Godwit

Marbled Godwits are among the larger and more common resident shorebirds in Saskatchewan. Throughout much of the southern half of the province, we see them wading in shallow marshes using their long bills to probe in deep mud for food, although they also wander nearby prairie uplands to feast on grasshoppers and bugs. Marbled Godwits often nest in pastures or fields a considerable distance from water. The shrill piercing call of these noisy birds makes them fairly easy to find.

Walking Tall

While most shorebirds have long legs, the Black-necked Stilt's legs are extraordinary, as if the bird is walking on...well, stilts. This shorebird, with its shocking pink legs, is not common in Saskatchewan, so finding it is always special. The limit of its summer range has traditionally been the northern United States, but more recently small numbers have made their way into Saskatchewan.

ABOVE: *Black-necked Stilt.*
RIGHT: *Male and female Wilson's Phalaropes.*
OPPOSITE: *Dowitchers.*

Whirling Waders

Brightly-coloured Wilson's Phalaropes travel all the way from their wintering grounds in southern South America to nest in marshes and sloughs through much of southern Saskatchewan. They differ from most birds in that the mates' roles are reversed. The much showier female (on the right) initiates the courtship, while the drab-looking male incubates the eggs and looks after the chicks. Feeding in shallow water, phalaropes resemble spinning tops as they twirl around in tight circles to churn up insects and other tasty morsels that they pick up with their sharp bills. They often hang out with avocets, taking advantage of any goodies that the larger birds may have stirred up with their feet.

Dowitchers

Feeding dowitchers are like a bunch of tiny sewing machines, rapidly and continuously jabbing their needle-sharp bills into shallow water or mud with such vigour that their heads are often a blur. Both Short-billed and Long-billed Dowitchers pass through Saskatchewan in spring and fall. Short-bills nest in far northern Saskatchewan and adjoining provinces and territories, while Long-bills head to the Arctic coast. Despite the names, the two are almost identical and used to be considered a single species.

Sanderlings

Sanderlings nest in the high Arctic and winter from the southern United States to South America, making refueling stops in Saskatchewan crucial. These little shorebirds often appear here in impressive numbers. It's been estimated that over half of the world's population stops at Chaplin Lake, where on one spring day over 55,000 were estimated to be resting and feeding. As suggested by the name, Sanderlings like beaches. We see them sticking close to the shoreline and dodging waves. They are fascinating to watch in flight, as large densely-grouped flocks dart about in unison, frequently and abruptly changing direction with ballet-like precision. Since the undersides of their wings and bodies are lighter-coloured than the upper sides, a magical change between light and dark occurs when they suddenly switch direction.

OPPOSITE AND ABOVE: *Sanderlings at Chaplin Lake.*

The Scoop on Pelicans

Enormous and magnificent, with a wingspan of up to three metres or 10 feet, American White Pelicans are our largest bird. Half of Canada's and close to a quarter of the world's population nests in Saskatchewan. Some major nesting sites include Prince Albert National Park (one of the largest colonies in Canada), Last Mountain Lake, and Redberry Lake—the centerpiece of Saskatchewan's first World Biosphere Reserve. Since pelicans travel long distances to feed, we might see them almost anywhere in the southern two-thirds of the province.

Pelicans are also a success story. Their dwindling numbers in the 1970s brought COSEWIC (Committee on the Status of Endangered Wildlife in Canada) to list them as a threatened species in 1978. But protection and a turn-around in their populations resulted in pelicans becoming the first bird species to be "delisted" by COSEWIC.

Saskatchewan abounds with marvelous places to watch pelicans, the easiest being below structures with frothing water, such as the weir on the South Saskatchewan River near downtown Saskatoon. Among our favourite spots is Last Mountain Lake. Here we can usually get quite close to where they congregate below the water control structures in the National Wildlife Area. Another hotspot often boasting impressive numbers is the spillway on Rafferty Dam near Estevan.

While man-made structures make it easy to observe pelicans, nothing compares with watching them feed along river rapids in the wilderness. Several prime spots are along the Churchill River, a "drop-pool" waterway where several lakes are linked together by sudden drops over rapids and waterfalls—perfect places for pelicans to hang out and do some fishing. But if we had to pick the top spot—anywhere—for pelicans, it would undoubtedly be Scoop Rapids on the Sturgeon-Weir River, a spot on the legendary fur-trade waterway where the fishing was said to be so good that you simply scooped them up. Over two hundred years ago, explorer Peter Fidler commented on the large number of pelicans here, and they still keep coming back. So intent are the great white birds on feeding in the boiling whitewater that they ignore us sitting just above them on rocky outcroppings.

The pelican's pouch is tailor-made for scooping up food in the water. The protrusion or "horn" on top of the bill is present only during breeding season.

Pelicans coming in for a landing.

ABOVE: *Pelicans feeding at Scoop Rapids on the Sturgeon-Weir River.*
OPPOSITE: *Sandhill Cranes.*

Land of Living Skies

Each fall, Saskatchewan hosts one of nature's great spectacles. Saskatchewan is on the Central North America Flyway, a migration route that birds follow between northern nesting grounds and southern wintering grounds. While a variety of birds follow this highway in the sky, it's the mind-boggling numbers of Snow Geese, White-fronted Geese, Canada Geese, and Sandhill Cranes that use this route and stage here in the hundreds of thousands that is truly outstanding. Last Mountain Lake, for example, could have up to 50,000 cranes and 450,000 geese during peak migration. Miry Bay and Galloway Bay on the South Saskatchewan River average close to 300,000 White-fronted Geese. But in an exceptional year, such as 1995, White-fronted Geese were estimated at 748,000, or about two-thirds of the central flyway's geese for that year.

It's fairly easy to see impressive displays at key sites such as the Quill Lakes, Foam Lake, Last Mountain Lake, or vantage points along the South Saskatchewan River. Even large ponds can host thousands of birds. Another convenient spot is Luck Lake, where we can watch and photograph in different directions along the dyke road crossing the lake. Arriving in daytime, we may only see a few birds resting on the water and nearby shores. But wait until sunset. As the sky turns shades of gold and crimson, it comes alive as wave after wave of waterfowl return from feeding in nearby fields to roost on the shallow lake. The noise is deafening as tens of thousands of voices resonate in the crisp fall air. At sun up, it gets even better. When the birds head off for another day of feeding, thousands upon thousands may lift off in unison, almost blackening the sky in a breathtaking display.

OPPOSITE: *Morning lift-off of geese.*
ABOVE: *Sandhill Cranes wade in the shallow water of Luck Lake as thousands of waterfowl return at sunset.*

Warblers

More than 30 different warbler species have been recorded in Saskatchewan. Most prefer mixed wood forests where they live almost completely on insects. Two species we commonly see include the eye-popping Yellow Warbler, and the more subtly-coloured Yellow-rumped Warbler which is the first to arrive in spring and the last to leave in fall. Unlike most warblers that stick to the forest belt and parts of the parklands, Yellow Warblers nest throughout Saskatchewan in forests, shrubs, farm yards, shelterbelts, and trees in urban areas.

OPPOSITE: *Male Yellow Warbler.*
TOP AND RIGHT: *Female Yellow-rumped Warbler.*
BOTTOM: *Male Yellow-rumped Warbler.*

Spring is Here

If we had to pick one sound that announces spring on the prairies, it would be the cheery song of the Western Meadowlark. When it seems as if the prairie is taking its time awakening from a long winter sleep, the early arrival of these bright yellow-breasted birds and their optimistic refrain reminds us that another cycle of life is beginning. Their wide-ranging song repertoire includes calls to attract mates and to out-do rivals. Meadowlarks are among the best-known birds on the open prairies of Saskatchewan; we frequently see them singing from fence posts right along the road. Unfortunately, their numbers are dwindling in places where intensive cultivation has left little or no grassy areas.

Baltimore Oriole

A sudden flash of orange high in the branches announces the male Baltimore Oriole. Unlike some birds, the female is colourful as well, but more yellow-olive, with slight tinges of orange. Spread throughout much of southern Saskatchewan, orioles normally stay high in the tree canopy where they nest and feed on insects. They are most active and easiest to see in spring when trees just start leafing out. While they like tall trees, orioles prefer semi-open woodlands such as farm yards and shelterbelts over thick forest.

LEFT: *Western Meadowlark.*
ABOVE: *Male Baltimore Oriole.*
OPPOSITE: *Male Mountain Bluebirds.*

Mountain Bluebird

Their flash of blue is eye-catching anytime, but is especially welcome when they arrive in early spring and add a splash of colour to drab vegetation that is only starting to show signs of life. Mountain Bluebirds are by far our most common bluebird, nesting in edges of woodlands, and on the prairie where nesting boxes are available. While tree cavities are their natural nesting sites, they also do well in artificial nesting boxes. Despite being widespread, Mountain Bluebirds are not especially plentiful in Saskatchewan, so seeing them is always a treat.

Bohemian Imbibers

Bohemian Waxwings are among our most sleek and elegant birds. Males and females are equally colourful and difficult to tell apart. They do have one failing, however; a tendency to fly "under the influence".

While some nest in the boreal forest and forest fringe of northern Saskatchewan, many of the birds we see here nest in the mountains. They travel through parts of central and southern Saskatchewan in winter and early spring in large flocks that often descend en masse upon a berry patch or fruit tree in a feeding frenzy.

We have an ornamental apple tree in our back yard in Saskatoon that attracts a variety of birds that nibble away at the fruit throughout the year. But when a hungry flock of waxwings swoops down, they strip almost every remaining apple from the tree. The problem is that by this time, the fruit has fermented, causing the waxwings to become a bit tipsy and to fly without due care and attention. Occasionally, one takes a notion to fly through a window, so we have to keep the blinds down almost constantly to discourage wavering waxwings from taking a shortcut.

ABOVE AND OPPOSITE: *Bohemian Waxwing.*

LEFT: *Chipping Sparrow.*
TOP: *White-throated Sparrow.*
BOTTOM: *Clay-colored Sparrow.*
OPPOSITE: *American Tree Sparrow.*

Sparrows

Over 20 different sparrow species have been recorded in Saskatchewan; some are distinctive while many have only subtle differences in their plumage. Sporting a rufous-red crown and prominent white "eyebrows", the Chipping Sparrow is named for the high-pitched "chip" sound of its call. While most common in forests where it prefers nesting in conifer trees, it's found through much of Saskatchewan, and has adapted well to farm yards and urban areas.

The American Tree Sparrow also has a red crown and other similarities to the Chipping Sparrow, but it has a bi-coloured bill and, just to be sure there's no confusion, an isolated dark breast spot that stands out as an "identification badge". In most of Saskatchewan, we see American Tree Sparrows during migration; they are usually the first sparrows to arrive in spring. Most nest in the far north and Arctic, and some winter in southern parts of Saskatchewan. Despite the name, these tree sparrows nest on the ground.

Less distinctive-looking Clay-colored Sparrows nest in the prairie and parklands, although they extend into parts of the north as well. Often seen foraging on the ground for seeds and insects, they especially like habitat with wolf willow, snowberry or other shrubs.

Whenever we hear the melodious call of a White-throated Sparrow, it immediately brings back memories of canoe trips on northern rivers, where a song that somewhat resembles "O Canada, Canada, Canada" breaks the stillness of the forest. While it also nests in wooded areas of central Saskatchewan and migrates through the south, this is primarily a bird of the north. Often we'll hear a White-throated Sparrow without ever seeing it, but when we do get a glimpse, we find an attractive bird with a distinctive white throat patch, bold yellow over the eyes, and black and white striped head.

Flashy Finches

These colourful seed-eating birds are noted for their stout bills, tailor-made for cracking seeds. The Purple Finch stands out as the flashiest looking. Famous bird guide author, Roger Tory Peterson, summarized its appearance as a "sparrow dipped in raspberry juice". Mostly northern birds, Purple Finches like the boreal forest and forest fringe, although we also see them in southern Saskatchewan during migration.

Handsome yellow American Goldfinches, on the other hand, nest roughly in the southern half of Saskatchewan, where they prefer semi-open areas. We often see them in disturbed areas, flying in their distinctive bobbing up and down manner and feeding on thistles, their favourite food. Both male and female are yellow, although the male's colouring is much bolder and he has a black "cap".

The House Finch has a unique history. Now common in Saskatchewan, the first one wasn't even seen in this area until the 1970s. Native to North America west of the Rockies, House Finches used to be captured and exported for the caged bird trade. Then in the 1940s an illegally-caught shipment was released in the New York area, and they eventually spread throughout the continent. Not many years ago, sightings in Saskatchewan were only occasional. Now these adaptable birds stay throughout the winter and are among the most gregarious visitors to feeders.

OPPOSITE: *Male Purple Finch.*
ABOVE: *Male American Goldfinch.*
RIGHT TOP AND BOTTOM: *Male House Finch feeding a chick.*

Rose-breasted Grosbeak

The blazing red triangular breast patch of the male is its most distinguishing feature, along with a sturdy bill that looks like it could crack open almost anything. Their song somewhat resembles that of robins. Rose-breasted Grosbeaks nest mostly in Saskatchewan's parkland and southern boreal forest. We often see these members of the finch family in southern parts of the province during migration.

Rose-breasted Grosbeaks.

Boisterous in Blue

Blue Jays are among the most easily recognizable as well as attractive birds, with blue and white plumage, a black "necklace", and prominent crest that can resemble a spiky hair-do. Mostly birds of the forest and parkland, they spread through much of the prairies as trees were planted after settlement. Blue Jays tend to be shy and secretive during nesting. We see them more readily in fall and winter when the few that over-winter here frequent feeders, often intimidating other birds. Blue Jays like to announce their arrival with a shrill call of "jaaay".

Blue Jays.

Western Kingbird

Western Kingbirds are flycatchers that expertly nab insects in mid-air. These fearless birds, known for driving away crows, hawks and other intruders, well deserve their "king" name. Historically, their main Saskatchewan haunt was the Cypress Hills and wooded areas with large trees along creeks and rivers in the southern part of the province. But with prairie settlement and more trees in farm yards and shelterbelts, Western Kingbirds have spread throughout much of southern Saskatchewan.

ABOVE: *Western Kingbird.* OPPOSITE: *Common Tern.* INSET: *Arctic Tern.*

Aerial Acrobats

Terns are remarkable birds to watch, as they have the uncanny ability to briefly hover in the air just before they dive into the water like torpedos to nab a small fish or minnow. Thanks to their angular wings and strongly forked tail, these acrobats of the sky can abruptly switch direction in flight. As suggested by its name, the Common Tern (left) is the most plentiful and widespread tern in Saskatchewan, found in most parts of the province with food-rich water.

Its similar-looking cousin, the Arctic Tern (inset), only nests in Saskatchewan's far north which is the southern edge of its range for inland Canada. Here it especially likes the sandy and gravelly plains of the Athabasca Sand Dunes on the south shore of Lake Athabasca. Hiking around these dunes, we could always tell if we had inadvertently strayed too close to tern nests, and we learned the hard way how aggressively they defend their nests and young. All of a sudden, screaming white and black missiles swooped from the sky and dive-bombed us repeatedly until we moved far enough away. Arctic Terns stand in a class by themselves when it comes to long-distance migration. They fly from Arctic Canada to as far south as Antarctica; no other creature on Earth travels that far.

These two terns bear a strong resemblance to each other, although the Arctic Tern has longer tail feathers and the male's bill is solid red, whereas the Common Tern has a black-tipped bill.

Northern Icon

Majestically perched atop a spruce tree, or soaring effortlessly overhead, the Bald Eagle is a symbol of the north. Saskatchewan boasts the largest inland concentration of Bald Eagles in the world. These huge birds are known for their enormous nests, often added to year after year, some as much as two metres across. While most nest in trees, eagles in Canadian shield country sometimes build nests on large rocks on the water's edge, a housing style we often see along Lake Athabasca's rugged north shore. Others prefer cliff-face notches, such as at aptly named Eagle Rock Lake on the Grease River. Numerous northern rivers including the Churchill and Clearwater, and lakes such as Besnard and Nemeiben attract fish-loving eagles. Our personal favourite eagle hotspot is the mouth of the MacFarlane River near Lake Athabasca where the mighty waterway spreads into wide shallow rapids teeming with fish. We've watched over 20 eagles at once soaring over the high sandy cliffs lining the water, with even more white-headed sentinels perched on overhanging jack pine boughs watching for their next meal.

OPPOSITE: *Bald Eagle flying.* INSET: *Bald Eagle chick in nest.*
ABOVE: *Skull Canyon on the Clearwater River is also known as Bald Eagle Gorge, named for eagles that sometimes nest on the craggy cliffs.*

ABOVE: *70-Mile Butte in Grasslands National Park. The open prairie of southern Saskatchewan provides prime hunting territory for hawks.*
OPPOSITE: *Swainson's Hawk.*

Swainson's Hawk

Swainson's Hawks are most common in the open prairie of southern Saskatchewan where they hunt for their favourite food—Richardson's ground squirrels, also known as gophers. These are Saskatchewan's most well-travelled hawks, migrating thousands of kilometres to their wintering grounds in the South American pampas. They watch for prey while soaring, although we often see them perched in trees or atop roadside fence posts. They tend to be one of the more "approachable" hawks, less likely to fly off the post immediately when we stop by the side of the road to have a closer look or take a photograph.

Nature's Cleaning Crew

They're as ugly as sin and stink to high heaven, but in the air, Turkey Vultures are among the most graceful fliers around, effortlessly soaring in the air currents with their outstretched wings spanning up to two metres. They stay aloft for hours, constantly scanning the ground for their favourite food—practically any dead critter. These carrion eaters play a crucial role as nature's cleaning crew.

Turkey Vultures inhabit roughly the southern half of Saskatchewan. While they aren't plentiful, their numbers have been increasing. Dr. Stuart Houston, a noted researcher who has studied and tagged Turkey Vultures for years, found that a significant reason for their success is that they have broadened their nesting options to include old abandoned buildings in rural areas, of which there are many. This has expanded their range north into the parklands.

ABOVE AND OPPOSITE: *Turkey Vultures.*

Great Horned Owl

This is our most common owl, and one we might find almost anywhere in Saskatchewan. While it prefers the parklands and southern boreal forest, it is also widespread throughout the south where it nests in wooded coulees, farm shelterbelts, and almost anywhere with trees. It certainly is "great"—one of our largest owls (only the Snowy Owl is slightly larger), with a wing-span as wide as 1.5 metres or around five feet. It doesn't have real horns, but rather feathery tufts that give the impression of horns. This is one of the earliest birds to nest in Saskatchewan, laying eggs in March or sometimes even late February. It's not like they're in a hurry to go anywhere; Great Horned Owls stay here year-round.

RIGHT: *Great Horned Owl.*
OPPOSITE: *Burrowing Owl.*

Burrowing Owl

The Burrowing Owl is in serious trouble, and is on the endangered species list in Canada. It is restricted to southern Saskatchewan and Alberta, the northern edge of its range. We often see this small prairie bird walking or standing on the ground rather than flying, since it nests in abandoned burrows of ground squirrels, prairie dogs, or badgers. One unique defence is a warning call to predators that mimics a rattlesnake. Some of its last strongholds remain in and around Grasslands National Park, near the South Saskatchewan River in southwest Saskatchewan, and in the Regina–Weyburn area where diminishing native grasslands have forced it to nest in ditches or other small grassy areas, and even in crop land.

Knock on Wood

Primarily woodpeckers of the forest, Yellow-bellied Sapsuckers are also common in the parklands, and bush areas in parts of southern Saskatchewan. Despite the name, they don't actually suck sap. Rather, they drill holes in trees, then when the sap begins to flow they lap it up with their brush-like tongues. Any tasty bugs sticking to the sap come as an added bonus. A tell-tale sign that sapsuckers are around is finding a series of precision-spaced holes drilled in a tree trunk or thick branch. Not that we generally need anything visual to know that sapsuckers are around; they like to announce their arrival with a series of squealing sounds that are almost like the meow of a cat. Of course, they tend to hammer away constantly: to feed, excavate homes, proclaim their territories, or attract mates. Sometimes they hammer on harder things than trees, such as power poles or even metal, just for the impressive sound it makes. Sapsuckers are common in the bush around our farm, so it's not unusual to be rudely awakened at 4:30 am when a sapsucker decides to demonstrate its hammering ability on the metal eavestrough—often right above the bedroom window.

OPPOSITE: *Yellow-bellied Sapsucker at nest hole.*
ABOVE: *Yellow-bellied Sapsucker cleaning its nest.*
RIGHT: *Male Northern Flicker, Yellow-shafted variety.*

Northern Flicker

The Northern Flicker is the most distinctive of our woodpeckers, as well as one of the most attractive. The Yellow-shafted variety of Northern Flicker is the most widespread throughout Saskatchewan, distinguished by its spotted front, bold red patch on the back of the head, black "moustache", and yellow undersides of wings. Compared to other woodpeckers, the flicker spends more time foraging on the ground for insects.

Hummers

Female Ruby-throated Hummingbird.

Male Ruby-throated Hummingbird.

Our smallest bird also ranks as one of the most beautiful. The Ruby-throated Hummingbird is named for the male's dark red, metallic-looking throat patch. Not much bigger than a large insect, this tiny bird migrates all the way from its wintering grounds in Mexico and Central America, and is capable of flying non-stop over the 800 kilometre expanse of the Gulf of Mexico. It is the only hummingbird that nests in Saskatchewan. We find it primarily in the southern forest, the parklands, and Cypress Hills, with fewer numbers in open prairie.

Hummingbirds always seems hyperactive; they constantly dart about, abruptly change direction, hover for a second or two, then shoot off like bullets. They have devised a series of manoeuvres as mating rituals or as intimidation to discourage rivals. The ones that nest near our farm yard seem especially partial to the "flying U", where they fly swiftly in a wide U-shape, the sound made by their wings rapidly changing pitch with the ups and downs—a performance that we usually hear before we see it. Hummingbirds can become reasonably comfortable around people; we could be two metres or less from our feeders, and they will still zoom in to feed. If one of us goes outside wearing a red hat, it's usually not long before we hear a "vroom, vroom" only inches away as an inquisitive hummer bolts in to investigate this odd-looking "flower".

Splendour in the Grass

The ultimate in resplendence, the male Ring-necked Pheasant is embellished with an iridescent blue-green head and neck, fire-engine red face, turkey-like wattles, formal-looking white collar, feathers with a complexity of chestnuts and coppers, and ending with a long elegant tail. Originally from Asia, pheasants aren't native here, but since they were introduced over a century ago, they have adapted well to many parts of central North America. In Saskatchewan, we see them most often in southernmost areas of the province. We came across this dapper male crossing a road just outside Estevan.

Male Ring-necked Pheasant.

Looking Sharp

Sharp-tailed grouse enjoy a special status as Saskatchewan's official bird emblem. Most common in south and central parts of the province, especially in pastures, Sharp-tails also occur farther north in open forest or along drier edges of bogs. For most of the year, males and females look similar and not especially colourful. But come spring mating season, all that changes.

In April, when the prairie slowly awakens from a long winter, the Sharp-tail lek, usually situated on a grassy knoll, is already alive with feverish activity. Males come to strut their stuff on these dancing grounds while females come to check out the guys. The normally lacklustre males puff themselves up, inflating bright purple air sacs on their necks, their yellow eye-combs fluffed up like bushy blonde eyebrows. Wings are drooped down, heads are lowered to almost ground level, and the name-sake pointed tail reaches skyward. Rapid-fire foot stomping, audible a kilometre away on a calm morning, propels them back and forth, round and round, like a bunch of out of control wind-up toys. Then as quickly as the dancing starts, it suddenly stops. Often two rival males end up face to face, staring each other down in a stand-off, waiting to see who blinks first. Occasionally these stand-offs lead to a fight, but usually they suddenly start dancing again, trying even harder to out-do competitors. Females watch from the sidelines or wander through the lek, searching for the most impressive mate. Watching this age-old ritual ranks among springtime's most unforgettable experiences.

OPPOSITE AND ABOVE: *Male Sharp-tailed Grouse in courtship display.*

Bushland Drummer Boys

Ruffed Grouse inhabit mixed wood forest throughout Saskatchewan, and wooded river valleys in prairie regions, but their favourite area is the aspen parkland. We might catch a glimpse of these masters of camouflage, then the next instant they just melt into the bush.

In spring and early summer, we're more likely to hear rather than see them. While they strut around and spread out their neck and tail feathers, males have an additional technique to both attract mates and proclaim territories—drumming. On a quiet morning we can hear them up to a kilometre away; a "put, put, put" sound, slow to begin with, then rapidly accelerating, and ending in a "prrrrr". The sound has been described as being like a motorboat, but for those of us who grew up in rural Saskatchewan, it's even more like a grain auger engine having trouble getting started. Males cup their wings and flap them wildly. While it looks as if they are thumping their chests, the sound comes from beating the air.

The moss-covered trunk of a fallen tree makes a perfect drumming log, and there could be more than one in different parts of the same bush. During one spring at our farm, two males had claimed drumming logs less than 50 metres apart. It was like dueling drums as they tried to out-do each other. But there was no real contest—one was obviously the better, and certainly the loudest drummer, while the other was the novice, perhaps a younger male still learning the ropes. Like other grouse, mating rituals move into full swing in the morning, although with Ruffies it's not unusual to hear drum rolls at any time of day.

LEFT AND OPPOSITE: *Male Ruffed Grouse.*

Sagebrush Strutters

We would be hard pressed to find a more pompous showoff than the male Greater Sage-Grouse. These guys really go all out to impress the gals, taking strutting and chest-puffing to a new level. Well before sunrise, we hear strange "plop, plop" sounds drifting through the crisp spring air. With dawn's first light, we finally make them out amongst the sagebrush, with tail feathers spread into magnificent fans and chests puffed to where they look like they'll burst. But looking good is only the start. A pouch in the male's esophagus is filled with air, while his head is tucked so deep into the neck feathers that it appears to be completely gone, giving a whole new meaning to "losing your head over a girl". When he suddenly releases the air, the skin collapses and produces the plopping sound, as the head pops back up like a jack-in-the-box. Sadly, this fascinating rite of spring is getting harder to see. The Greater Sage-Grouse is not only Canada's largest grouse, but also the rarest. On the endangered species list, this bird is in serious trouble; the few remaining in Canada are restricted to a small area of southwest Saskatchewan and southeast Alberta. Among its last strongholds is the short grass prairie in and around Grasslands National Park. As its name suggests, the sage-grouse depends on sagebrush; it's the main part of the bird's diet in summer and it lives on it almost exclusively in the winter. If the sagebrush disappears, the sage-grouse will too.

ABOVE AND OPPOSITE: *Greater Sage-Grouse.*

White-tailed Deer

Graceful and fast, white-tailed deer dart with short galloping spurts, their name-sake tails upright and waving like a flag. Faced with a fence, they simply leap over with little apparent effort. White-tailed deer are considered the most abundant large herbivore in North America. Saskatchewan goes a step farther in not only having a lot of white-tails, but more deer with large antlers than anywhere on the continent, including some world records. They are widespread in the southern half of the province, especially where the agricultural belt mixes with bushland, although some venture farther north into the boreal forest.

The white-tail has long enjoyed a special status in Saskatchewan: it is our official animal emblem, and is featured on the province's Coat of Arms.

ABOVE AND RIGHT: *White-tailed deer fawn.*
OPPOSITE: *White-tailed deer buck.*

Mule Deer

Named for their large mule-like ears, mule deer also have a white rear-end, but they don't wave their tails like white-tailed deer. They are among the most fascinating animals to watch as they run. At high speed, they don't so much run as bounce, as if their stiff-looking legs were spring loaded. Slightly larger than white-tailed deer, mulies tend to be less skittish when we encounter them in the wild, especially in places such as Grasslands National Park where they are used to people. These are mostly deer of the grasslands and shrublands of the southwest, although in recent years they have spread throughout much of southern Saskatchewan, and even into parts of the north.

OPPOSITE: *Mule deer doe.* ABOVE: *Mule deer bucks.*

Speed Demons of the Prairie

Named for prongs or points on their horns, pronghorn are unique to the North American Great Plains, and range from northern Mexico to southern Saskatchewan and Alberta. Suited to wide open spaces, pronghorn are the fastest land animals in North America, capable of reaching 100 km/hour. Worldwide, the pronghorn's speed comes second only to that of the cheetah. Pronghorn inhabit southern parts of Saskatchewan, and are especially prevalent in the southwest in places such as Cypress Hills, Grasslands National Park, Coteau Hills, and the Lake Diefenbaker area. In recent years they have expanded their range, and now we see them farther north. From the tens of millions that roamed the grasslands before settlement, they were almost decimated in the early 20th century due to pressures of settlement and over-hunting. But careful management has brought them back from the brink to where their numbers have remained reasonably stable.

OPPOSITE AND ABOVE: *Pronghorn.*

Elk.

Elk

This large and beautiful red deer is most famous for the majestic antlers sported by the males. Elk are the most vocal of the deer family, especially during fall rutting season when the male's call starts as a low roar then blends into a higher pitched bugling sound that reverberates through the forest.

Once common on the prairies, elk in Saskatchewan live mostly in the northern agricultural belt and forest fringe where they graze and browse on woody vegetation. The Porcupine Hills near Hudson Bay are especially known for a high elk population, but the easiest place to find and watch elk is Prince Albert National Park where they have become accustomed to people. The name of the park's townsite, Waskesiu, is derived from a Cree word meaning red deer, or elk. Pockets of elk are also scattered throughout the south, and their numbers are increasing in many prairie areas. We once came across a small herd of elk stags, all with magnificent antlers, in open crop land near Lake Diefenbaker, something we found quite exciting but which local residents didn't consider unusual.

Moose

Moose are the largest members of the deer family. Their long legs seem out of proportion to their bodies, but are tailor-made for moose to wade and munch water lilies and other aquatic food, or to straddle small trees that they bend over to reach the twigs and leaves.

Encountering moose in the wild is always a highlight of our northern canoe trips. They are the least sociable deer, usually found alone or as a cow-calf pair. Moose are generally shy and don't like to be detected, although in remote areas where there is little hunting pressure, they may continue feeding and seem unconcerned if we don't approach too closely or make noise.

Surprising moose is a different story; usually they run, but not always. On one of our northern canoe trips, an ill-tempered bull moose charged us when we rounded a sharp creek bend and suddenly came eye to eye with him.

Moose sightings are becoming more common in southern Saskatchewan. They live in forested areas such as the Cypress Hills, Moose Mountain uplands, and bushlands along the South Saskatchewan River near the Red Deer River forks. We even encounter moose in open grasslands and crop land.

Bull moose browsing on leaves.

Tree Huggers

It's called a black bear, but its colour can range from black to brown to cinnamon to almost white. Seeing bears in the wilderness is always a thrill, yet this is the animal that people tend to worry about most when travelling in the north. Unfortunate encounters do indeed happen, though we have never had a serious bear incident over several years of canoeing and camping in northern Saskatchewan. But that doesn't mean we can be foolhardy. While camping on the shore of the William River, we saw an enormous black bear amble out of the forest towards us. We managed to scare it back into the bush, but it turned slowly and grudgingly, as if to say "Okay, I'll go...for now." That was our cue to pack up and find a new camp spot. The most bizarre encounter we've had was canoeing in a swift current on a remote northern river, when a bear came out of the bush and decided to swim across the river directly in front of us, without bothering to check for traffic. Just before we were about to crash into it, it glanced up and its eyes almost popped out of its head as it did a quick flip and swam madly back to shore.

We find bears almost anywhere in the north and occasionally in wooded areas farther south. Some places are noted for healthy populations such as the south shore of Lake Athabasca or Selwyn Lake. To look for bears, any of the northern or forested parks are good bets including Prince Albert National Park, Meadow Lake Provincial Park, or Duck Mountain Provincial Park.

LEFT AND OPPOSITE: *Not all black bears are black.*

Dogs of the Wild

Red fox.

Of all the dog-like carnivores, the wolf is the most powerful symbol of the northern wilderness, its haunting howl on a quiet evening sending shivers down the spine. This arch predator of the north hunts all hooved animals as well as small critters; a pack of wolves can take down a moose, and even large bears will shy away from a wolf pack.

We might find wolves almost anywhere in northern Saskatchewan, although they tend to be secretive. On northern canoe trips, we frequently come across wolf tracks on sandy or muddy shores, yet only occasionally do we ever get a glimpse of them. Where sandy beaches are common on rivers and lakes such as the Fond du Lac River, William River, or south shore of Lake Athabasca, the number of tracks can give the impression that the countryside is practically crawling with wolves.

Coyotes are much easier to find. While most common on the prairies, they live throughout Saskatchewan except in the extreme far north. The most numerous of the dog-like carnivores, their numbers have been increasing over the years. Although coyotes are plentiful, chances are that we'll hear them more than see them. In many rural areas, their loud yip-yipping calls are almost a nightly ritual. They play a useful role in helping to control small rodents. Coyotes have adapted well to people, but this close proximity often leads to conflicts.

The most elegant of our dog-like animals, red foxes are sleek and graceful, with pointed nose and pointed ears. Their trademark, however, is their long luxurious tail, stretching almost as long as their body. While found throughout Saskatchewan, foxes prefer open areas and edges of woodlands over dense forest. They are very adaptable and seem reasonably comfortable living close to people. It's not unusual to find fox dens in farm yards or even on the edges of cities.

TOP: *Wolves.*
BOTTOM: *Coyote.*

ABOVE: *American badger.* OPPOSITE: *Richardson's ground squirrels.*

One of the larger members of the weasel family, American badgers inhabit the prairie anc parkland regions of Saskatchewan, although their numbers have been decreasing throughout their range. Highly efficient digging machines, they excavate impressively large holes for burrows and to go after prey. Among their favourite foods are Richardson's ground squirrels, also called gophers. Usually considered a nuisance by farmers, these ground squirrels are common throughout the prairie and parklands of Saskatchewan. Coyotes, foxes, and several different hawks also prey on ground squirrels, helping to control their numbers.

Black-tailed Prairie Dog

Saskatchewan is the only place in Canada that is home to black-tailed prairie dogs, and the only place in Saskatchewan to see them is along the Frenchman River valley in and around Grasslands National Park. Here they occupy the northern tip of their range that stretches south along the Great Plains to northern Mexico.

Driving along the valley road through Grasslands National Park, it's easy to assume that these prairie dogs are thriving, judging by the seemingly countless burrows as far as we can see. These social animals live and work together in colonies to raise their young, and protect the colony using several different warning screams. They are easy and entertaining to watch since many burrows line the road, and the dogs normally tolerate people who stay in their cars. Constantly scurrying about, they pop in and out of the burrows, make a variety of chattering and chuckling sounds, or let out a sharp warning scream if they sense danger. Watch for the unusual "jump-yip" display where a prairie dog jumps into the air while making dog-like yipping sounds, thought to be an "all-clear" signal to the colony.

In some ways, prairie dogs resemble ubiquitous Richardson's ground squirrels, better known as gophers, found throughout southern Saskatchewan. However, they are considerably bigger than gophers and have distinctive black-tipped tails.

RIGHT: *Black-tailed prairie dog keeping a lookout.*
OPPOSITE TOP LEFT: *Family greeting at a prairie dog burrow.*
OPPOSITE TOP RIGHT: *Prairie dog getting ready for a "jump-yip".*
OPPOSITE BOTTOM: *Prairie dog picking up lunch.*

Flamboyance in Flight

When we look for beauty and colour in nature, butterflies are hard to beat. They pack a dazzling range of hues and intricate designs into their small sizes. Saskatchewan has records of 160 species. We find them in varied habitats from woodlands to open prairie, roadsides, back yard gardens, and practically anywhere with flowering plants. The best thing about watching butterflies is that we don't have to go far to find them and they often let us get up close and personal.

OPPOSITE: *Fritillary.*
TOP LEFT: *Tiger swallowtail.*
TOP RIGHT: *White admiral.*
BOTTOM RIGHT: *Northern crescents.*

Carpets of Colour

Saskatchewan's diverse ecosystems make for a wildflower extravaganza, with different regions of the province having their own specialties. If we had to pick one part of Saskatchewan for wildflowers, it would be the Cypress Hills. Not only do we find grassland and forest species, but also plants more common in alpine altitudes. We can return throughout the summer and be rewarded with different displays. Late May and early June are prime times for orchids, whose number and diversity cannot be matched anywhere on the prairies, while mid-summer's splendid array of grassland flowers blankets the hillsides.

Any place with native prairie is prime flower habitat, such as Grasslands National Park, the parks and natural areas around Lake Diefenbaker, and pastures almost anywhere. Traditionally dry areas of the southwest such as near the South Saskatchewan River and Red Deer River forks can also be rewarding. When conditions are right, a carpet of magenta covers the ground with low pincushion cactus blossoms, and at other times it's a vivid yellow with prickly pear cactus.

Another of our favourites is Duck Mountain Provincial Park, where mixed woods combine with a southern boreal forest. Early summer in this upland environment brings flashy displays of yellowy-orange hoary puccoons, blue lungwort, and showy marsh marigolds along creek beds.

The north also has its share of specialties, such as pinkish-red stemless lady's-slippers, expanses of yellow sand heather on sandy areas, or hardy plants that manage to survive in impossibly small cracks in rocks in the Canadian shield.

OPPOSITE: *Summer tapestry of wildflowers in the Cypress Hills highlighted by bright yellow gaillardia in the foreground.*
ABOVE: *Prairie rose.*
RIGHT: *Prairie coneflower.*

LEFT: *In areas with poor moisture conditions such as eroded hillsides or clay slopes, smooth blue beardtongue adds a welcome touch of colour.*
RIGHT: *Pinedrops grow on a long stalk and have delicate bell-shaped flowers, but no leaves or chlorophyll. In Saskatchewan, pinedrops are rare, and are found in only a few locations such as the lodgepole pine forest of the Cypress Hills.*

OPPOSITE: *Wildflowers in the Cypress Hills including wild bergamot (purple) and cinquefoil (yellow) in the foreground.*

Flower clusters of spreading dogbane, found along roadsides and in wooded areas, look like brightly striped candy canes. Toxins in a milky sap in its leaves and stem help protect it from browsing animals.

Western red lily, the floral emblem of Saskatchewan. Lilies grow in many different habitats including prairie meadows, aspen and boreal forests, hillsides and slopes, and in disturbed areas such as ditches. They generally bloom in late June and early July.

ABOVE: *Sticky geranium (left) and mountain shooting star (right) are both specialties of the Cypress Hills, adding splashes of pink and purple to grassy meadows.*
BELOW LEFT: *Delicate flowers of hoary puccoon appear in early summer in moist meadows and edges of woodlands, mostly in central and eastern Saskatchewan.*
OPPOSITE: *High plains of southern and southwest Saskatchewan are home to bluish-purple silvery lupines, growing here along with cinquefoil.*

LEFT: *The aptly named three-flowered avens with nodding reddish flowers that grow in groups of three is common in open prairie and parkland meadows.*
ABOVE: *A harbinger of spring, the prairie crocus blossoms so early that there may still be some snow on the ground. Most common in grasslands, especially native prairie, the crocus also grows in some open areas of the parklands.*
OPPOSITE: *Heart-leaved arnica with its cheery sunshine face is most prevalent in higher altitude coniferous forests such as in the Cypress Hills.*

Cactus Country

The yellow-flowered prickly pear cactus (above left) and the magenta-flowered pincushion cactus (above right) add colour to the often dry slopes of the South Saskatchewan River valley. The view opposite is overlooking the South Saskatchewan River "Loop" where the waterway flows into Saskatchewan. After only a few kilometres, the river loops back into Alberta, then seems to change its mind yet again as it winds back into Saskatchewan to continue its eastward journey. This beautiful valley was in danger of being flooded by the proposed Meridian Dam to be built where the river crosses the Fourth Meridian, the border between Saskatchewan and Alberta. The project was shelved as uneconomic, giving the valley a reprieve, at least for now.

Orchids

Saskatchewan has 28 species of orchids found anywhere from dry coniferous forests to moist bogs or exposed rocky slopes. All are delicate plants that are sensitive to disturbance.

The showy calypso orchid, also known as venus slipper, (left) grows in the lodgepole pine forest in Cypress Hills, as does striped coralroot (opposite right) which has no chlorophyll and no roots, but derives its nutrients from decayed plant matter. The pinkish-red stemless lady's-slipper (opposite top) prefers dry areas in the northern boreal forest. It is especially prolific in the bushes bordering the Athabasca Sand Dunes, where in one place there were so many that we had to be careful where we stepped. More widespread throughout Saskatchewan in moist forests, coniferous bogs, or even roadside ditches is the yellow lady's-slipper (opposite bottom) with its brilliant lip or pouch.

Living by Water

OPPOSITE: *Marsh marigolds thrive in wet areas such as swamps, moist meadows, and along streams.*

ABOVE: *A thick patch of swamp horsetail grows around a rock along Besnard Creek. Horsetails are common in wet meadows, marshes, lake edges, and along creek beds, especially in the boreal forest.*

While we usually associate fall colours with the changing of the leaves, in Saskatchewan we enjoy the added dimension of changing colours on the prairie as well as in the forest.

It's often in transition areas that we find the most magnificent displays, where golden deciduous trees of the aspen parkland gradually blend into the evergreen spruce and pine of the boreal forest. Hotspots include Prince Albert National Park, Narrow Hills Provincial Park, Meadow Lake Provincial Park, or Duck Mountain Provincial Park, to mention only a few. Farther north, the amber colour often comes from birch trees scattered through the boreal forest, or from tamaracks whose needles turn yellow in fall and drop off.

We can also head south in search of fall colour. Well-treed areas such as Moose Mountain Provincial Park, the Cypress Hills, or parts of the Qu'Appelle Valley provide an attractive mix of colourful leaves and changing grassland vegetation. Prairie coulees with their diverse grasses, crimson-leaved shrubs and scattered trees can make for some outstanding sights. Among our favourite places in fall is around the forks of the Red Deer River and South Saskatchewan River west of Leader. Nearby Chesterfield Flats has one of the province's highest concentrations of cottonwood trees that turn a brilliant gold in fall.

ABOVE: *South Saskatchewan River near the Red Deer River forks.*
OPPOSITE: *Mixed forest in Narrow Hills Provincial Park.*

To see colour in the forest, we need to look down as well as up.

OPPOSITE: *Fallen leaves combine with forest floor vegetation for a kaleidoscope of colour in Prince Albert National Park.*
ABOVE: *Beauty can be deceiving; the amanita mushroom is attractive though poisonous. This specimen grew in our farm yard. One fall day we watched as a red squirrel broke it into pieces and carried it away.*
TOP RIGHT: *Dry-ground cranberry growing in northern Saskatchewan's Canadian shield.*
BOTTOM RIGHT: *Bunchberry leaves in transition as they gradually change from green to deep crimson in fall.*

The colours of fall come in many shades. The brilliant red of mountain maples in the Pasquia Hills north of Hudson Bay (above) contrasts with the rich golden hues of a birch tree at Corner Rapids on the Churchill River (opposite).

Southern Saskatchewan has its own fall colour highlights.

ABOVE: *The Qu'Appelle Valley at wildlife-rich Nicolle Flats in Buffalo Pound Provincial Park.*
LEFT: *Cottonwood trees at Chesterfield Flats along the South Saskatchewan River near Estuary.*
OPPOSITE: *Brunyee Coulee in Saskatchewan Landing Provincial Park.*

Resources

To learn more about Saskatchewan's wildlife, and the many opportunities for wildlife viewing and photography, contact the following key resources:

Nature Saskatchewan
www.naturesask.ca   1-800-667-4668
Nature Saskatchewan is dedicated to conserving the province's natural environment. The organization has outreach programs, does research, publishes a variety of nature-related books, and has meets with popular field trips. The website has links to local nature societies and affiliated organizations throughout Saskatchewan that also operate numerous programs and field trips.

Saskatchewan Provincial Parks
www.saskparks.net   1-800-205-7070
Some of the top wildlife viewing possibilities are in our provincial parks, which cover diverse landscapes and habitats throughout Saskatchewan. The parks website links to pages with further details on each park.

Tourism Saskatchewan
www.sasktourism.com   1-877-2ESCAPE
The provincial tourism promotion organization provides travel guides, maps and information on all parts of the province, including nature-oriented destinations, and tour operators offering trips.

Ducks Unlimited Canada
www.ducks.ca
DUC is dedicated to conserving wetlands in Canada, and is a prime source of information on waterfowl. Of special interest is their series of Nature Watch driving tour brochures to several wetlands around Saskatchewan.

Canadian Wildlife Service
www.cws-scf.ec.gc.ca
Extensive information on wildlife in Canada. The agency is also responsible for Migratory Bird Sanctuaries and National Wildlife Areas such as Last Mountain Lake.

Parks Canada
www.pc.gc.ca
Links to further information on Saskatchewan's two national parks: Prince Albert National Park, and Grasslands National Park.

In addition to facilities in provincial and national parks, Saskatchewan has numerous nature interpretive centres. The following is a small sample:

Chaplin Nature Centre, Hwy#1, Chaplin
www.chaplintourism.com   1-877-746-7273

Saskatchewan Burrowing Owl Interpretive Centre, Moose Jaw
www.sboic.ca   306-692-8710

Quill Lakes International Bird Area Visitor Centres
Wynyard, Wadena, Foam Lake
www.quill-lakes-bird-area.com

Beaver Creek Conservation Area, near Saskatoon
www.meewasin.com

Last Mountain Lake National Wildlife Area
306-836-2022

Last Mountain Bird Observatory
www.naturesask.ca

Old Man on His Back Interpretive Centre, Southwest of Eastend
www.natureconservancy.ca

Redberry Lake Biosphere Reserve
www.redberrylake.ca

Other books by Robin and Arlene Karpan

Saskatchewan Scenic Secrets
Take a spectacular visual journey through Saskatchewan's natural landscapes, from border to border to border to border. More than stunning photography, the informative text explains the significance of Saskatchewan's many special landscapes.
Full colour, hard cover • 11 x 8 1/2 , 128 pages • 145 colour photos
2 colour maps • ISBN 978-0-9683579-3-4 • $34.95

Northern Saskatchewan Canoe Country
Discover the north's most spectacular rivers and lakes. This coffee-table book features 230 stunning images complemented by stories on the rich history, the tremendous diversity of landscapes, and the Karpans' personal experiences in the north. Winner of Tourism Media Award for Travel Journalism, Saskatchewan Tourism Awards of Excellence.
Full colour, hard cover • 11 x 8 1/2 , 128 pages • 230 colour photos
ISBN 978-0-9809419-0-6 • $34.95

Northern Sandscapes – Exploring Saskatchewan's Athabasca Sand Dunes
The Athabasca Sand Dunes are like nowhere else on Earth – a desert-like environment seemingly misplaced in the northern forest. Follow the Karpans on an exciting canoe journey through this stunning wilderness. Winner of Souvenir of the Year Award, Saskatchewan Tourism Awards of Excellence.
Soft cover, perfect bound • 8 1/2 x 11, 128 pages • Index
115 colour photos • ISBN 978-0-9683579-0-3 • $29.95

Larger Than Life – Saskatchewan's Big Roadside Monuments
Know where to find the world's largest still for making home brew? How about the world's largest moose, bunnock, or white-tailed deer? Why did a Saskatchewan town build a huge Santa Claus, and others build a big pink pig, a lighthouse, or kangaroo rats? Travel to over 70 communities to marvel at massive monuments, and discover the stories behind who built these intriguing sculptures, and why.
Soft cover, perfect bound • 5 1/2 x 8 1/2, 176 pages • 100 b&w photos
ISBN 978-0-9683579-9-6 • $18.95

Saskatchewan Scenic Drives
Head off the major highways and into Saskatchewan's scenic heartland. Follow secondary highways, grid roads, and backroads to spectacular river valleys, forested lakelands, wild badlands, secret hideaways, and awesome vistas. Drive a third of the way across Saskatchewan entirely in the Qu'Appelle Valley. See the other-worldly "Crooked Trees". Precise directions and maps make it easy to discover these and many more Saskatchewan scenic wonders.
Soft cover, perfect bound • 6 x 9, 240 pages • 45 b&w maps
120 b&w photos • ISBN 978-0-9683579-6-5 • $24.95

Saskatchewan Trivia Challenge
This book is packed with over 240 intriguing questions on important (and completely frivolous) tidbits of information on Saskatchewan's biggest, best, firsts, records, famous folks, oddities, and just plain "cool stuff". Winner of Souvenir of the Year Award, Saskatchewan Tourism Awards of Excellence.
Soft cover, perfect bound • 5 1/2 x 8 1/2, 192 pages • 85 b&w photos
ISBN 978-0-9683579-2-7 • $14.95

Western Canadian Farm Trivia Challenge
It's time to have some fun with farming. Which country is the biggest export market for our wheat? How much does a cow pie weigh? How many seeds are there in a bushel of canola? Which community built a statue to honour a pig that escaped from the abattoir?
Soft cover, perfect bound • 5 1/2 x 8 1/2, 192 pages • 66 b&w photos
ISBN 978-0-9683579-5-8 • $14.95

Saskatoon History Trivia Quest
Discover Saskatoon's fascinating past – famous and colourful characters, accomplishments, the flamboyant boosterism of the boom years, and tidbits of oddball stuff that contributed to the development of Saskatchewan's largest city.
Soft cover, perfect bound • 5 1/2 x 8 1/2, 176 pages • 100 b&w photos
ISBN 978-0-9683579-8-9 • $16.95

Available at fine bookstores, or from Parkland Publishing. Phone: 306-242-7731 www.parklandpublishing.com

FINAL FRAME: *Pelican in flight.*